Modeling Rapidly Composable, Heterogeneous, and Fractionated Forces

Findings on Mosaic Warfare from an Agent-Based Model

TIMOTHY R. GULDEN, JONATHAN LAMB, JEFF HAGEN,
NICHOLAS A. O'DONOUGHUE

Prepared for the Defense Advanced Research Projects Agency
Approved for public release; distribution unlimited

NATIONAL DEFENSE RESEARCH INSTITUTE

For more information on this publication, visit www.rand.org/t/RR4396

Library of Congress Cataloging-in-Publication Data is available for this publication.
ISBN: 978-1-9774-0447-3

Published by the RAND Corporation, Santa Monica, Calif.
© Copyright 2021 RAND Corporation
RAND® is a registered trademark.

Cover photos: *rancho_runner/Getty Images/iStockphoto;*
alxpin/Getty Images; Terryfic3D/Getty Images.

Cover design: Rick Penn-Kraus

Support RAND
Make a tax-deductible charitable contribution at
www.rand.org/giving/contribute

www.rand.org

Preface

As the U.S. Department of Defense transitions from a focus on irregular warfare to great-power competition, several new approaches to fighting conflicts are under consideration to reduce costs and increase effectiveness and robustness. These approaches have emerged as complements or even alternatives to a more traditional focus on high-capability, high-cost platforms, such as the F-35 fighter, B-21 bomber, or *Ford*-class aircraft carrier.

One new approach, championed primarily by the Defense Advanced Research Projects Agency (DARPA), is known as *Mosaic warfare*, after the idea of creating a complex image out of many small, simple pieces. This approach relies on fractionation of capabilities from large multicapability platforms onto multiple smaller ones, the ability to employ heterogenous mixes of capabilities throughout a battlespace, and finally, the ability to rapidly compose a set of needed capabilities in a time and place to accomplish a mission. DARPA is exploring many possible solutions to various issues with Mosaic warfare through a series of internal projects, such as Collaborative Operations in Denied Environment, OFFensive Swarm-Enabled Tactics, Adapting Cross-domain Kill-webs, Air Combat Evolution, System of Systems Integration Technology and Experimentation, Cross-Domain Maritime Surveillance and Targeting, and Assault Breaker II. However, these projects are primarily focused on addressing implementation challenges (such as ensuring successful command and control of assets or closing a kill chain) without addressing a more-fundamental question: Is Mosaic warfare better (i.e., more cost-effective and robust) than more-traditional approaches?

In 2019, DARPA asked the RAND Corporation's National Security Research Division (NSRD) to explore and validate the fundamental value propositions of the three key Mosaic warfare architectural attributes—*fractionation, heterogeneity,* and *dynamic composition*—by means of reduced-order modeling and simulation. The study also aimed to identify near-term Mosaic experimentation opportunities and experimental hypotheses through functional decomposition and recomposition of existing systems and architectures. The research project, *Mosaic Warfare Experimentation Architecture Study,* was conducted in NSRD's Acquisition and Technology Policy Center. The results of this project are documented in this report and in the following RAND reports:

- Justin Grana, Jonathan Lamb, and Nicholas A. O'Donoughue, *Findings on Mosaic Warfare from a Colonel Blotto Game,* Santa Monica, Calif.: RAND Corporation, RR-4397-OSD, 2021.
- Nicholas A. O'Donoughue, Samantha McBirney and Brian Persons, *Distributed Kill Chains: Drawing Insights for Mosaic Warfare from the Immune System and from the Navy,* Santa Monica, Calif.: RAND Corporation, RR-A573-1, 2021.

This report is focused on an agent-based modeling approach, one of the two modeling efforts undertaken by the project.

This research was sponsored by DARPA and conducted within the Acquisition and Technology Policy Center of the RAND National Security Research Division (NSRD), which operates the National Defense Research Institute (NDRI), a federally funded research and development center sponsored by the Office of the Secretary of Defense, the Joint Staff, the Unified Combatant Commands, the Navy, the Marine Corps, the defense agencies, and the defense intelligence enterprise.

For more information on the RAND Acquisition and Technology Policy Center, see www.rand.org/nsrd/atp or contact the director (contact information is provided on the webpage).

Contents

Figures and Tables

Figures

Tables

Summary

As the U.S. Department of Defense transitions from a focus on irregular warfare to great-power competition, several new approaches to conflict are under consideration to reduce costs and increase effectiveness and robustness in the face of evolving adversary technologies and tactics. These approaches have emerged as complements or even alternatives to a more traditional focus on high-capability, high-cost platforms, such as the F-35 fighter, B-21 bomber, or *Ford*-class aircraft carrier. One new approach, championed primarily by the Defense Advanced Research Projects Agency (DARPA), is known as *Mosaic warfare*, after the idea of creating a complex image out of many small, simple pieces.[1] This approach relies on fractionation of capabilities from large multicapability platforms onto multiple smaller ones, the ability to employ heterogenous mixes of capabilities throughout a battlespace, and the ability to rapidly compose a set of needed capabilities in a time and place to accomplish a mission.

This report presents an intentionally abstract representation of a variably Mosaic architecture that is nominally based on air operations but is designed to elucidate some fundamental principles of all such systems and begin to address fundamental questions about when Mosaic architectures perform better (i.e., are more effective and more robust) than more-traditional multicapability systems.

---

[1] Defense Advanced Research Projects Agency, "DARPA Tiles Together a Vision of Mosaic Warfare," webpage, undated.

Agent-Based Modeling

To explore the potential benefits of the Mosaic approach, we have developed an agent-based model that tasks a set of airframes to service a set of targets. The model is implemented on the NetLogo agent-based modeling platform. Each platform has a fixed number of these capabilities on board, chosen from among six possible capability types. A monolithic airframe would possess all six. A fully fractionated platform would possess only one. At levels of fractionation between these, capabilities are assigned to airframes at random without replacement. Therefore, a given platform with three capabilities might have capabilities 0, 2 and 5, while a second one might have capabilities 1, 2 and 3.

To be serviced, a target needs to be visited by a collection of platforms that have the required capabilities. The complexity of target demands can vary independently from the fractionation of airframes. Targets can require anything from all six capabilities to a single capability for servicing. The complexity and number of both airframes and targets can be varied across model runs.

Measures of Effectiveness

We calculate a basic measure of effectiveness as the *number of capabilities delivered against targets per minute*. This metric tells us how efficiently the platform capabilities are being used and is invariant to target complexity. This metric is computed by summing all the capabilities delivered in the model run to that point and dividing by the number of simulated minutes. The simulation is run until this metric stabilizes.

Orchestration of Airframes

A key goal of our study was to explore several variations on the set of rules that control the behavior of airframes in prioritizing targets. These rules are designed to provide reasonable performance in matching capabilities to targets, but do not aim to be optimal, as optimality in this area is not realistically achievable. Rather, the object of these heuristics is to show some of the difficulties presented by the design of orchestration rules in even this minimally complex environment. The most basic rule is a purely greedy algorithm, with two optional subordinate rules that aim to avoid some of the worst inefficiencies (assigning a distant platform to a target when a closer one can address it, or assigning redundant platforms to targets).

Observations

We find that a Mosaic approach can be particularly effective against highly fractionated targets that do not require large assemblies of simple platforms to service targets. Mosaic platforms have the potential to perform as well as monolithic platforms if efficient teams can be created and maintained when appropriate, but suboptimal teaming rules can easily produce performance degradations of as much as 50 percent.

Mosaic architectures can perform better than more-monolithic architectures across a broad range of target demands, but only with high-performing, robust orchestration rules. We find that the best choice of rules is highly dependent on both platform and target factors. This dependence is difficult to predict in advance of an operation.

We conclude that DARPA would be wise to invest in the development of teaming and orchestration strategies that are dynamic and adaptive, so that Mosaic platforms are afforded the ability to respond to changes and uncertainty in the environment by altering how they coordinate their activities.

Acknowledgments

The authors would like to express their sincere thanks to Dr. Timothy Grayson, Director of the Defense Advanced Research Projects Agency's Strategic Technology Office (STO), and LtCol Daniel Javorsek, U.S. Air Force, a Program Manager in STO and our action officer, for their insight, direction, and support. We would also like to extend our appreciation to Ronald Hill and David Ott, support contractors for Dr. Grayson and LtCol Javorsek, respectively, for their endless assistance coordinating the details of this project. We wish to thank Samuel Earp and John Kamp, who provide subject matter expertise to STO and were instrumental in strengthening our analysis.

During the review and quality assurance process, this report evolved greatly. The authors wish to thank John Murphy, Aaron Frank, Jim Powers, and Joel Predd for their assistance reviewing the documents.

Abbreviations

AFSIM	Advanced Framework for Simulation, Integration, and Modeling
C2	command and control
DARPA	Defense Advanced Research Projects Agency
DoD	U.S. Department of Defense
MoE	measure of effectiveness
TSP	Traveling Salesman Problem

CHAPTER ONE

Introduction

As the U.S. Department of Defense (DoD) transitions from a focus on irregular warfare to great-power competition, several new approaches to fighting conflicts are under consideration to reduce costs and increase effectiveness and robustness in the face of evolving adversary technologies and tactics. These approaches have emerged as complements or even alternatives to a more traditional focus on high-capability, high-cost platforms, such as the F-35 fighter, B-21 bomber, or *Ford*-class aircraft carrier. One new approach, supported by the U.S. Army and U.S. Air Force, is known as *multidomain operations* and looks to build nonlinear "kill webs" by rapidly bringing together sensors and shooters across service, domain, and functional stovepipes.[1]

Another new approach, championed primarily by the Defense Advanced Research Projects Agency (DARPA), is known as *Mosaic warfare*, after the idea of creating a complex image out of many small, simple pieces.[2] This approach relies on fractionation of capabilities from large multicapability platforms onto multiple smaller ones, the ability to employ heterogenous mixes of capabilities throughout a bat-

---

[1] For example, see the remarks from ADM Philip Davidson, Commander of the U.S. Indo-Pacific Command, before the Senate Armed Services Committee in early 2019, in which he said: "[T]he U.S. government must continue to pursue multidomain capabilities to counter anti-air capabilities . . . ." (U.S. Senate, Senate Armed Services Committee, "Statement of Admiral Philip S. Davidson, U.S. Navy Commander, U.S. Indo-Pacific Command, Before the Senate Armed Services Committee on U.S. Indo-Pacific Command Posture," Washington, D.C., February 12, 2019).

[2] Defense Advanced Research Projects Agency, "DARPA Tiles Together a Vision of Mosaic Warfare," webpage, undated.

1

tlespace, and the ability to rapidly compose a set of needed capabilities in a time and place to accomplish a mission.

The Mosaic approach to warfare is generally defined as a shift from multicapability systems toward dynamically composed collections of more-specialized systems.[3] The scope of the Mosaic concept is potentially extremely broad, covering a range of land, sea, air, space, and cyber operations and the effectiveness of the approach is likely to vary tremendously depending on the details of any given system and situation. This report presents an intentionally abstract representation of a variably Mosaic architecture that is nominally based on air operations but is designed to elucidate some fundamental principles of all such systems and begin to address fundamental questions about when Mosaic architectures perform better (i.e., are more effective and more robust) than more-traditional multicapability systems.

We conceive of a Mosaic system as having three primary characteristics: fractionation, heterogeneity, and rapid composability. Fractionation varies from *monolithic* (in which all capabilities required to achieve a mission reside in a single platform)[4] to *highly fractionated* (in which those capabilities are spread across a large number of platforms). This is a somewhat inherent characteristic of the platform itself, although individual capabilities on a monolithic platform could be fractionated out. An F-35 might be thought of as monolithic while a collection of a dozen drones that work together to achieve the same mission would be highly fractionated. If the radar on an F-35 was briefly used by someone other than the pilot, that could be thought of as a temporarily fractionated capability. Heterogeneity varies from *monocultural* (in which all the platforms employed have the same mix of capabilities) to *highly heterogeneous* (in which platforms have non-

---

[3] David Deptula, Heather Penney, Lawrence Stutzriem, and Mark Ginzinger, *Restoring America's Military Competitiveness: Mosaic Warfare*, Arlington, Va.: Mitchell Institute for Aerospace Studies, September 2019.

[4] We are referring to the definition of *monolithic* in terms of a system or organization, which is "large, powerful, and intractably indivisible and uniform" (Lexico.com, "Monolithic," webpage, undated). This is admittedly problematic, in that the systems are not indivisible per se, but they are indivisible insofar as they assign all capabilities together, whereas a Mosaic arrangement is free to assign the capabilities separately.

overlapping capability sets). Finally, we define *orchestration* as the rules or algorithms that control how platforms coordinate and deconflict their activities to accomplish assigned missions. Many levels of orchestration would be needed, from simple autopilots keeping platforms in the air to theaterwide command and control (C2) guidance assigning platforms to geographic areas and operations. Here, we primarily refer to orchestration at a mission- or package-planning level: Given a specific demand, what capabilities and platforms are best suited to meet that demand and how should they operate to do so?

Optimal orchestration in any but the most trivial cases is computationally nonachievable—which, in computer science, is referred to as an *NP-hard problem*.[5] Rather than developing a highly refined rule set to deal with our abstract case, or attempting to systematically survey the space of possible rules (which is extremely large), we developed a set of fairly simple heuristics for our modeling effort that produce reasonable performance over various situations and that provide a test bed for developing a general understanding of the monolithic-versus-Mosaic trade space.

Even in a highly simplified and abstract space, the fitness landscape for orchestration rules can become quite complex, with small shifts in the environment or in the rules producing large changes in system performance that can be hard to predict in advance of actual simulation. For example, *teaming*, in which groups of capabilities are formed and kept together for a period of time, can be a critical part of

---

[5] E. L. Lawler, J. K. Lenstra, A. H. G. Rinnooy Kan, and D. B. Shmoys, *The Traveling Salesman Problem: A Guided Tour of Combinatorial Optimization*, Hoboken, N.J.: John Wiley and Sons, 1985. The model presented here is closely related to a classic problem in computer science that is called the Traveling Salesman Problem (TSP), which is the most accessible instance in the class of NP-hard problems. NP is an abbreviation for "nondeterministic polynomial time." The basic TSP involves finding the shortest path to visit a series of points in a plane without revisiting any of them, and it is notoriously difficult to solve in a provably optimal way—although heuristic rules can be used to produce relatively good solutions quickly. Our platforms are essentially doing this but with complications: Each time a target is visited, it is removed and replaced with a new target in a random location. This has the effect of presenting a new TSP each time a target is visited. The problem becomes even more complex when multiple platforms with multiple overlapping capabilities are introduced.

system performance in the Mosaic context. However, knowing when teams should be formed and for how long are challenges to simple orchestration rules. As a result of these issues, we explore a variety of orchestration rules for performance and robustness as key variables in the following sections.

Caveats and Limitations

It should be noted that this model seeks to elucidate some general properties of the Mosaic warfare approach but not to simulate any particular implemented system. For this reason, the model has been kept at an abstract level—representing only as much of a variably Mosaic system as is absolutely required to produce insight into the general problem. We avoided the temptation to include low-fidelity representations of realistic issues, because we believe that this would obscure the insights that arise from an abstract model while not actually providing enough fidelity to evaluate the added capabilities. For example, we did not include the effects of adversary jamming or C2 complexity on the effectiveness of a Mosaic system. Similarly, we assumed that all platforms travel at the same speed and have not included attrition (because of reliability issues or enemy action) as a characteristic of any of the platforms. These realistic factors are omitted, not because they are difficult to represent, but because the fitness space became quite complex when the model included only different levels of fractionation under different target conditions using different orchestration rules.

Every aspect of this model is ripe for future study and further analysis. This includes the need for a detailed trade space study of orchestration rules, decision logic for when to employ each set of rules and repetition of results with more-realistic capabilities (including some capabilities that can be employed at standoff and others that have a limited magazine), target-servicing logic (e.g., a kill chain that requires capabilities be delivered in order), and diversity in platforms (such as a correlation between the number of capabilities carried and the platform speed). Additionally, a critical feature of Mosaic warfare is likely to be its resilience to attrition and platform failures because of the disaggregated nature of its capabilities.[6] An extension of this work with threats to the aircraft, or random dropouts, could be enlightening.

There is a limit to what can be done in a reduced-order model, however. Detailed analysis on specific postulated or realized architectures may necessitate the use of a more complex modeling and simulation capability. Notably, the Advanced Framework for Simulation, Integration and Modeling (AFSIM) is an alternative mission-level simulator that represents individual agents with various decision algorithms and is equipped with a number of realistic sensor and kinematic performance models.[7] The AFSIM is not the only applicable tool in existence, but it enjoys wide adoption across DoD, particularly within Air Force research communities. Choosing the appropriate modeling environment (reduced order or more detailed) will be an essential first step in future projects in this area.

---

[6]   Deptula et al., 2019.

[7]   Peter D. Clive, Jeffrey A. Johnson, Michael J. Moss, James M. Zeh, Brian M. Birkmire, and Douglas D. Hodson, "Advanced Framework for Simulation, Integration and Modeling (AFSIM)," in *Proceedings from International Conference on Scientific Computing (CSC '15)*, Las Vegas, July 2015.

CHAPTER TWO

Agent-Based Modeling Approach

To explore the potential benefits of the Mosaic warfare approach, we have developed an agent-based model that tasks a set of platforms to service a set of targets. The model is implemented on the NetLogo agent-based modeling platform, which provides a flexible development environment, ease of model sharing, and adequate performance to test the resulting model over the course of thousands of runs.[1]

In the following examples in this chapter, we will consider architectures with six possible capabilities. Although these might be thought of as electro-optical/infrared sensor, radar, jammer, long-range communication device, munition A, and munition B, it is important to remember that we are abstracting from the details of these functions—each one is simply a function that may or may not be needed to accomplish a given mission. The model simply notes whether a capability is present or absent on each platform and whether target servicing requires a particular capability type.

Platform Fractionation

Each platform has a fixed number of these capabilities on board. A monolithic platform would possess all six. A fully fractionated platform would possess only one. At levels of fractionation between these, capabilities are assigned to platforms at random without replacement.

---

[1] U. Wilenski, *NetLogo*, software package, Center for Connected Learning and Computer-Based Modeling, Northwestern University, Evanston, Ill., 1999.

Therefore, a given platform with three capabilities might have capabilities 0, 2, and 5, while a second one might have capabilities 1, 2, and 3. Platforms cannot repeat a capability on board (i.e., a platform with capabilities 1, 1, and 2 is not allowed), but they might overlap across the force (e.g., two separate platforms can both carry capability 1). To keep our terms clear, we will refer to this spectrum as ranging from *monolithic* to *Mosaic*. We do not vary the level of heterogeneity of the architecture as an independent variable—instead, we hold it constant by controlling the number of capabilities present in the battlespace.

The goal of the platforms is to service targets. Platforms and targets interact in a 40 nmi × 40 nmi space with connected borders. This means that going off one side brings the platform back onto the other side and going off the top brings it back onto the bottom in a manner that is common in arcade games. Platforms move at a nominal speed of 400 knots (nmi/hour) when they are not waiting at a target. They are initially positioned in random locations with random headings.

Target Complexity

To be serviced, a target needs to be visited by a collection of platforms that have the required capabilities. The complexity of target demands can vary independently from the fractionation of platforms. Targets can require anything from all six capabilities to a single capability for servicing. To differentiate from platforms, we will refer to this spectrum as moving from *complex* to *simple*. In the base model case, all platforms must arrive at the target before it can be serviced. Platforms arriving early loiter on the target until the full set has assembled unless they decide to leave the target to pursue another mission—in which case, their capabilities must be replaced at the target by those of other platforms before servicing can be completed.[2] In the base case, targets

---

[2]   We realize that, if implemented literally, loitering near the target would likely increase attrition. This loitering behavior is a simple and abstract way to represent things in our model. With proper coordination, all platforms could easily arrive simultaneously (with any required loitering happening in a safer place). Even the simultaneity is an abstraction here, as some capabilities might be required before others, or might be time independent. The

are replaced in a random location as soon as they are serviced—thus maintaining the general density of targets throughout the run.

The complexity and number of both platforms and targets can be varied across model runs. For the sake of comparability, we looked at sets of runs in which the total number of airborne capabilities available and the total number of capabilities needed by the targets are both held constant, with "Mosaicness" of the platforms and complexity of the targets being varied. For example, in a run with a total of 120 capabilities available, and six capability types, the model would have 20 platforms in the monolithic case (6 capabilities × 20 platforms = 120 total) and 120 platforms in the fully Mosaic case (1 capability × 120 platforms = 120 total). Note that both of these extremes also have identical levels of heterogeneity. Similarly, if there are 120 total capabilities needed by the targets, there are 20 targets in the complex target case and 120 targets in the simplest case.

Measures of Effectiveness

We calculate a basic measure of effectiveness (MoE) as the *number of capabilities delivered against targets per minute.* This metric tells us how efficiently the platform capabilities are being used, and it is invariant to target complexity.[3] This metric is computed by summing all of the capabilities delivered in the model run to that point and dividing by the number of simulated minutes. The simulation is allowed to run until this metric stabilizes. In a run with six capabilities needed per target, if 100 targets had been serviced in ten minutes, the calcula-

---

requirement of simultaneous presence can be thought of as the simplest possible rule for coordinated action—a representation that is appropriate to the current abstract analysis.

[3]   Our original metric was targets serviced per time, but it was difficult to compare simple (i.e., easier) target demands with more-complex (i.e., difficult) ones because target complexity is a desired input variable. This earlier metric produced meaningful results when comparing different levels of Mosaicness in platforms at the same level of target fractionation but led to an apparent degradation of performance as one moved from simple targets requiring only one service to complex ones requiring as many as six. The current metric removes this predictable (and relatively uninteresting) trend—allowing more-meaningful comparisons across both platform Mosaic levels and target complexity levels.

tion would be 100 targets times six capabilities needed per target all divided by 10 minutes: $(100 \times 6)/10 = 60$ capabilities delivered per minute. Without considering the financial costs of integration versus fractionation (which is beyond our scope here), this seems the best way to make runs that can be meaningfully compared with one another as the Mosaicness of platforms and complexity of targets is being varied.

Because the targets are replenished (each target is replaced at a new position once it is addressed), there is no ending condition. The model is run until a steady-state is reached, where the MoE is stable. This is operationalized by allowing each model run to simulate 2,000 minutes of activity. At this point in the simulation, there is still some random variation in the metric, but that variation is small and is without significant trend. This also avoids the need to perform time-consuming Monte Carlo runs[4] despite the initial randomness.

Orchestration of Platforms

As noted earlier, we knew that orchestration rules could possibly dominate the effects of fractionation on performance. Thus, a key goal was to explore several variations on the set of heuristic rules that control the behavior of platforms in prioritizing targets. These heuristics are designed to provide reasonable performance in matching capabilities to targets but do not aim to be optimal, because optimality in this area is not realistically achievable. Rather, the object of these heuristics is to show some of the difficulties presented by the design of heuristics in even this minimally complex environment. The most basic heuristic rule is a purely greedy algorithm, and there are two optional rules. The three are explained as follows:

- **Basic Rule:** At each time step, all unassigned platforms attempt to select a target to address using the process outlined in Figure 2.1. This is done by identifying all targets that require one or more of

---

[4] *Monte Carlo runs* refers to repeated trials with randomized inputs. The results are then aggregated across the full set of randomized trials to increase statistical stability and confidence.

the capabilities that a platform possesses. Those targets that have inbound platforms to address all of their required capabilities are ignored. The closest remaining potential target is then chosen, and the platform begins to travel to the target. The platform loiters at the target until the target is serviced by platforms carrying all of its required capabilities. There is a provision to place priority on targets that are "stuck" (those that have some capabilities addressed but not all). When platforms having all of the target's needed capabilities have assembled at the target, it is considered "serviced." It is removed from the simulation and is replaced by another randomly generated target. All of the platforms that had been assigned to the now-serviced target begin traveling again and look for new targets to address.

- **Recursive Targeting Rule:** If the closest target to a platform requiring relevant services is already covered by another platform that is closer to the target, the first platform examines the next closest target, running through all targets to see if there is one that could use its services and is not already covered by a closer platform.
- **Check Needs Rule:** When platforms initially observe potential targets, the capabilities of all platforms already assigned are subtracted from the original target demands. This reduces redundant platforms assigned to the same target.

Figure 2.1 illustrates the Basic Rule Set governing the assignment of platforms to targets, including the resolution of a "stuck" target—one that is in progress but has not been fully serviced. The Recursive Targeting Rule and Check Needs Rule are not shown. In our tests, we compared the performance of the Basic Rule Set alone and with either or both of the modifications, for a total of four orchestration rule cases.

The specific details of how orchestration and targeting rules were implemented is beyond the scope of this report; our intention is to show the utility of reduced-order modeling and to make broad inferences about Mosaic warfare, where possible, rather than to produce objective, repeatable results.

Figure 2.1
Target Assignment and Deconfliction Approach

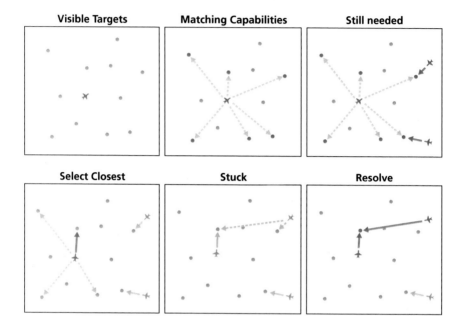

CHAPTER THREE

Results

In the previous chapter, we introduced our basic targeting heuristic and two enhancements that are intended to improve performance in certain complex scenarios. We ran all four of these targeting rules (Basic, Basic + Recursive Target, Basic + Check Needs, Basic + Both) through a series of test scenarios where we varied

- **target complexity,** from simple (single capability needed) to complex (all six capabilities needed)
- **platform complexity,** from simple (single capability) to complex (all six capabilities)
- **target density,** from a target-poor environment (1:1 ratio of targets to platforms) to a target-rich environment (4:1 ratio of targets to platforms)

We then posited an ideal decision engine that chose the best rule set for each scenario and plotted performance for this optimal rule set in each situation. These calculations represent an approximate lower bound on achievable performance. It is the best that is achievable with the four rule sets defined here but does not preclude the existence of higher-performing rule sets for each condition.

Table 3.1 summarizes the capabilities delivered per minute when we varied the complexity of the platforms and targets in a target-poor (1:1 ratio of targets to platforms) environment. For quick reference, the cells shaded with green represent greater than 300 capabilities delivered per minute, yellow cells represent greater than 200 capabilities per minute, and red cells represent less than 200 capabilities per minute.

Table 3.1
Performance (Capabilities Delivered per Minute) Trends in a Target-Poor Environment

Capabilities per Target	Capabilities per Platform					
	1	**2**	**3**	**4**	**5**	**6**
1	327	255	216	186	164	153
2	280	325	193	256	129	214
3	280	105	325	166	139	254
4	177	181	254	179	143	283
5	140	269	272	209	166	308
6	263	273	280	204	192	321

NOTE: Cell shading indicates an approximate level of performance. Green = greater than 300 capabilities per minute. Yellow = greater than 200 capabilities per minute. Red = less than 200 capabilities per minute.

The exact results are not as informative as the trends, given our reduced-order modeling. In general, we see that platforms with four or five capabilities perform poorly. This is likely due to the fact that there are six capabilities in the environment: They are unlikely to form effective teams (without needlessly duplicating some capabilities), while the one-, two-, three-, and six-capability platforms can easily do so when needed. Similarly, the four- and five-capability targets are the most difficult to address with teams; therefore, they generally fare worse and are both best serviced by fully capable platforms (six capabilities per platform).[1]

Table 3.2 repeats this analysis for the target-rich (4:1 ratio of targets to platforms) environment. This analysis is done by reducing the number of platforms; therefore, the absolute performance numbers are smaller and we set the highlight bounds at 60 and 90. In this case, we see again that the four- and five-capability platforms perform the

---

[1]   Consider, for example, a target that requires five capabilities. A nearby five-capability platform might not have the right mix (i.e., the target might require the one capability that the platform is missing), necessitating a second platform. This team would carry a total of ten capabilities, five of which would be wasted on the current target.

Table 3.2
Performance (Capabilities Delivered per Minute) Trends in a Target-Rich Environment

Capabilities per Target	Capabilities per Platform					
	1	2	3	4	5	6
1	101	72	58	49	43	40
2	100	102	48	70	33	58
3	99	47	103	49	40	73
4	49	51	83	58	46	83
5	38	91	93	67	50	93
6	98	100	100	72	63	101

NOTE: Cell shading indicates approximate level of performance. Green = greater than 90 capabilities per minute. Yellow = greater than 60 capabilities per minute. Red = less than 60 capabilities per minute.

worst, and that the four- and five-capability targets perform the worst, likely because of their indivisibility by 6 (the total number of capabilities) making it difficult to efficiently form teams or divide up targets.

To further illustrate trends, we plot the same data in Figure 3.1 with the capabilities per platform on the horizontal x-axis, and the performance on the vertical y-axis. The figure shows the performance of this metaheuristic for different levels of target complexity ranging from highly complex (six capabilities needed per target) as the solid blue line to very simple (one capability needed per target) as the solid red line. For each of these levels of complexity, the figure plots the performance of different levels of Mosaicness, ranging from monolithic (six capabilities per platform) on the left side of the figure to Mosaic (one capability per platform) on the right side of the figure.

Against the simplest targets (solid red line), we see a steady increase in performance as platforms become more Mosaic. This makes sense because five of the six capabilities of each monolithic platform are essentially "wasted" on a target that requires only one capability. Note, however, that the gain when moving from monolithic to Mosaic is not a factor of six but only a factor of 2.5. This is because the monolithic

Figure 3.1
Performance of Metaheuristic Across Mosaic Levels in Target-Rich Environment

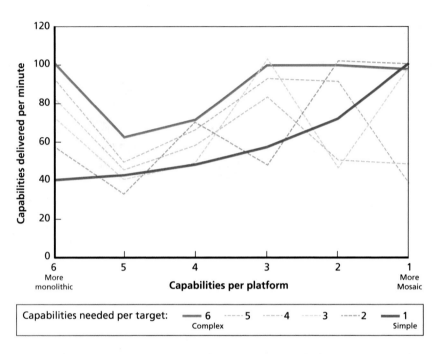

NOTE: In a target-rich environment, platforms provide a total of 60 capabilities and targets demand a total of 240 capabilities. It should be noted that the lines in this case are a visual convenience that allow the viewer to track the performance of the metaheuristic for a given environment across different levels of Mosaicness. They are not meant to imply that the performance spectrum is in any way continuous. As the current model is configured, there is no meaning to the idea of having 4.5 capabilities per platform. Although it is easy to imagine mixed setups in which a number like this could represent an average, the current model does not do this. There are a large number of ways to construct a mixed force, and exploring them would unmanageably expand the scope of the current report.

platform is able to proceed to any nearby target and service it, while a Mosaic platform needs to move to the nearest target that requires its particular capability, which is likely to be farther away. Therefore, the

Mosaic platforms need to cover more distance over time to apply their more-specific capabilities.[2]

Against complex targets (solid blue line), we see a nonmonotonic curve. The performance of the most monolithic and Mosaic platforms is quite similar, which stands to reason because the Mosaic platforms can always form up into teams of six and move together as a single team, servicing targets in exactly the same way that a monolithic platform would. We see similar performance when platforms have two or three capabilities, because these can similarly be formed into teams of three and two platforms respectively to cover all six capabilities—which are needed at each more-complex target. Platforms with four and five capabilities, however, are less efficient. Teams of two are still required to provide all six capabilities, but some capabilities will be redundant and therefore wasted. This would lead to a notable performance drop for the four-capability platforms and an even more significant drop for the five-capability platforms.

The dotted lines show performance against intermediate levels of target complexity. The dotted blue line represents targets requiring five capabilities. Performance across levels of fractionation in platforms is quite similar to fully complex targets, although a bit lower, except in the case of fully Mosaic platforms, which perform surprisingly poorly under the metaheuristic used here. We suspect that a more-refined algorithm could improve performance in this area. The dotted gray line represents less-complex targets requiring three capabilities. We see a similar degradation of performance across the more monolithic end of the platform range but sharp declines at both fully and mostly Mosaic platforms (i.e., having one and two capabilities, respectively). This is also likely illustrating the limit of the heuristic used here and is exacerbated by the large number of ways that six capabilities can be taken as sets of two, making it hard for the algorithm to intelligently

---

[2] We note that increased distance in a hostile environment would likely increase attrition, though differences in the vulnerabilities of different types of platforms make this difficult to model in a fully abstract way. This would be an interesting avenue for further research. Larger distances covered would also increase fuel consumption per target serviced, reducing the number of targets that might be serviced per sortie and increasing maintenance requirements.

assemble stable teams and causing a large number of long-distance and inefficient flights to reach targets.

Comparison Between Rule Sets

To understand the importance of orchestration, we now analyze which rule set provided each of the performance levels reported in Tables 3.1 and 3.2. Table 3.3 indicates the best-performing rule set in each scenario for a target-poor environment via both the text and cell shading, while Table 3.4 does the same in a target-rich environment.

The table is designed to highlight the way that different orchestration rules perform best in different parts of the Mosaic and fractionation space, with the areas of different colors having best performance from different rules. Looking at orchestration rule choice in this target-poor environment, the basic rules (yellow cells) are most effective when Mosaic platforms match up against more-complex targets, while the addition of the Recursive Targeting Rule (red) improves performance for more-monolithic platforms going against more-fractionated targets. However, neither of these rule packages perform well when moderately

Table 3.3
Heuristic Rule Set Comparison in a Target-Poor Environment

Capabilities per Target	Capabilities per Platform					
	1	2	3	4	5	6
1	BR + RT	BR + RT	BR + RT	BR + RT	BR + RT	BR + RT
2	BR	BR + RT	BR + RT	BR + RT	BR + RT	BR + RT
3	BR	BR	BR + RT	BR	BR	BR + RT
4	BR	BR	BR	BR	BR + RT	BR + RT
5	BR + RT	BR	BR	BR + CN	BR + CN	BR + RT
6	BR	BR	BR	BR + Both	BR + CN	BR + RT

NOTE: BR = Basic Rule Set (yellow). BR + RT = Basic Rule Set + Recursive Targeting (red). BR + CN = Basic Rule Set + Check Needs (blue). BR + Both = Basic Rule Set + Both Additional Rules (purple).

Table 3.4
Heuristic Rule Set Comparison in a Target-Rich Environment

Capabilities per Target	Capabilities per Platform					
	1	2	3	4	5	6
1	BR + RT	BR + RT	BR + RT	BR + CN	BR + Both	BR + CN
2	BR + RT	BR + RT	BR + RT	BR + RT	BR + RT	BR + RT
3	BR + Both	BR + RT	BR + RT	BR	BR + RT	BR + Both
4	BR + RT	BR + RT	BR + RT	BR + RT	BR + RT	BR + CN
5	BR + RT	BR + CN	BR + CN	BR + Both	BR + CN	BR + RT
6	BR + Both	BR + Both	BR + Both	BR + CN	BR + Both	BR + RT

NOTE: BR = Basic Rule Set (yellow). BR + RT = Basic Rule Set + Recursive Targeting (red). BR + CN = Basic Rule Set + Check Needs (blue). BR + Both = Basic Rule Set + Both Additional Rules (purple).

monolithic platforms (with four or five capabilities per platform) go up against moderately concentrated targets (with four or five capabilities needed per target), with significant gains coming from replacement of the Recursive Targeting Rule with the Check Needs Rule or using both additional rules together. We also note that the four-, five-, and six-capability platforms required four different rules for best performance, while the one-, two-, and three-capability platforms only needed two. This is at least some indication that architectures that are more Mosaic could be more robust to orchestration rule quality.

Table 3.4 shows the comparison of heuristics in a target-rich environment. In this case, we keep the number of required target capabilities constant at 240 but reduce the number of airborne capabilities to 60. This has the effect of increasing the ratio of targets to platforms, making the environment more target rich.

The target-rich case produces a major shift in orchestration rule choice, however, with the Basic Rule Set (yellow) providing best performance only in one mixed case of four capabilities per platform and three capabilities needed per target—and performance here is actually very similar to the Recursive Targeting Rule Set. The Recursive Targeting Rule set (red) dominates much of the space, with the Check

Needs Rule (blue and purple) becoming important with more-complex targets. While the combined set of all rules provides marginally superior performance in only a few situations, its performance is close to that of the best performing heuristic in all cases.

The robust good performance of the combined rule set for target-rich environments is in marked contrast to the target-poor case, where the combined rule set was distinctly inferior to simpler rules in many situations—providing only half the performance of the Basic Rule Set or Recursive Targeting Rule in many parts of the space.

Two key findings emerged from the analysis of the MoE for the target-poor and target-rich scenarios described in Tables 3.1 through 3.4:

1. Orchestration performance can vary greatly with small shifts in the target demand.
2. Rules can interact in ways that are hard to predict in advance, even in a highly simplified environment.

The rules for teaming in Mosaic configurations can be hard to adjust for particular situations—and poor teaming performance can seriously degrade overall system performance.

Variation Between Rule Sets

Although useful, the information in Tables 3.3 and 3.4 does not address the variation between the best and worst rule set in each case. Figure 3.2 plots the normalized spread, defined as the difference between the capabilities per minute for the best- and worst-performing rules in each case, divided by the best-performing rule. This metric is less than one by definition, but we find the range of values to be between .03 (there is only a 3-percent difference in performance) and 0.6 (there is a 60-percent difference in performance). The values are plotted in Figure 3.2 as a series of bar charts. For each target complexity case, we highlight which aircraft setting generated the largest spread with a red bar and note the normalized spread at the end of the row.

Figure 3.2
Normalized Spread Between Available Rule Sets

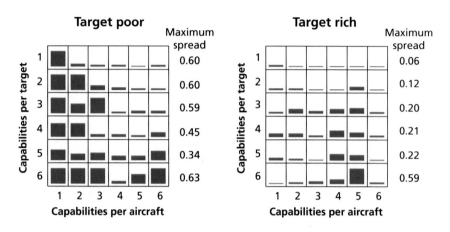

NOTE: Each table entry contains a vertical bar indicating the normalized spread of that case between the best-performing rule set and the worst-performing rule set, divided by the best-performing rule set, and is thus bounded between 0 and 1. Bars shown have a maximum height of 0.5. The peak spread in each row is indicated by a red bar, and its value is reported at the end of the row

We observe that, in general, the target-poor case exhibits greater variation among rule sets, with most of that variation occurring when the aircraft are Mosaic-like or the targets are very complex. In both target-rich and target-poor scenarios, there are cases where the choice of rule set seems to have little impact (normalized variation is less than 5 percent) and other cases where it has significant impact (normalized variation is greater than 50 percent).

This analysis only considers the four rule sets we devised; this is not a systematic survey of possible orchestration rules. However, we gathered from this analysis that the impact of a given orchestration does depend on the relative richness of targets and resources, the relative complexity of the aircraft, and the targets that they attempt to service. In the next section, we will analyze some of these relationships in greater detail.

Impact of Recursive Target Rule Depends on Target Complexity

In our modeling efforts, we found that target complexity and target density can play a critical role in best rule set choice. To illustrate this, Figure 3.3 plots the performance, measured by capabilities delivered per minute, as a function of the total number of platform capabilities (in all cases, the number of target capabilities required is held constant at 240). At the left edge of the x-axis, we have target-poor environments; at the right edge, we have target-rich environments. Performance of the Basic Rule Set is plotted with dotted lines, while the Basic + Recursive Rule Set is plotted with solid lines. Blue curves correspond to simple targets (each target requires one capability), and red curves correspond to complex targets (all six capabilities required). In all cases,

Figure 3.3
Impact of Platform Density on Performance

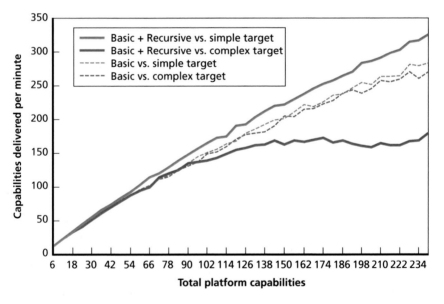

NOTE: All platforms are fully fractionated, meaning that they carry one capability per platform. The total number of target capabilities required is 240, leading to 40 complex targets or 240 simple targets.

the airborne platforms are fully fractionated, meaning that each platform carries only one capability.

The Basic Rule Set provides very similar performance against both concentrated (red dotted line) and fractionated (blue dotted line) targets and that performance generally increases as additional platforms are introduced. Adding the Recursive Targeting Rule, however, leads to a strong change in performance. It provides a meaningful advantage against simple targets (blue solid line), consistently outperforming the Basic Rule Set, but it produces a quite significant disadvantage against complex targets, and this disadvantage increases as the environment becomes more target-poor (red solid line moving to the right).

The cause of this performance breakdown is the Recursive Targeting Rule's tendency to break up teams in pursuit of higher utilization rates. For example, imagine that three platforms are assembled to service a target, but the next nearby target needs only two of them. With the Recursive Targeting Rule, the third platform can find a target where its capabilities are needed, and it breaks off to service that target. After the second target is serviced, the next target might require the capabilities of the first platform that broke off—but that platform is now far away, possibly working with another group. In a target-poor environment, this creates a great deal of churn among teams, leading to degraded performance overall as groups break up and reform in pursuit of individually high utilization. While a greater percentage of platforms have an assignment at any given time, they are spending more time in transit and end up serving fewer targets per minute.

Summary of Results

1. In our simple modeling environment, a Mosaic architecture can improve performance by a factor of more than two when servicing simple targets. This is achieved by having more platforms that can simultaneously address separate targets.
2. When targets are complex, Mosaic strategies can achieve parity with monolithic strategies if efficient teams can be created and maintained, but they can underperform by as much as 50 per-

cent if these teams cannot be efficiently assembled and maintained.

3. Alignment between the team capabilities employed and the true target demands is essential. Teams of monolithic and partially Mosaic platforms can be inefficient because they bring redundant capabilities to bear on a target.

4. Target density emerged as an important variable in orchestration rules, having a large impact on the relative performance of various orchestration heuristics.

CHAPTER FOUR

Observations

This report has explored a highly abstract model of Mosaic warfare operations that allows for the controlled exploration of three dimensions of variability: fractionation of platforms, complexity of targets, and the relative density of platforms and targets. We find that a Mosaic warfare approach can be particularly effective against highly fractionated targets that do not require large assemblies of simple platforms to service targets. Mosaic platforms have the potential to perform as well as monolithic platforms if efficient teams can be created and maintained when appropriate, but suboptimal teaming rules can easily produce performance degradations of as much as 50 percent. We found that alignment between team capabilities and target needs is critical and that a Mosaic force needs to be able to assemble "lean" teams that bring just the right mix of capabilities to each mission.

A key finding is the criticality of orchestration rules to the overall performance of Mosaic architectures. Mosaic architectures can perform better than more-monolithic architectures across a variety of target demands but only with high-performing, robust orchestration rules. We found that the best choice of rules is highly dependent on both platform and target factors and that this dependence is nonlinear and therefore hard to predict in advance of an operation. This has important implications for the design, test, and employment processes behind any Mosaic warfare system that DARPA might want to investigate or in which DoD might invest.

We have explored only the three factors discussed (fractionation of platforms, complexity of targets, and the relative density of platforms and targets) in a highly idealized environment. There are myriad other dimensions in which systems might vary, such as a variable number of capabilities per platform; variable capability performance (e.g., reliability, effectiveness, or range) on some platforms; variable platform performance (e.g., speed, maneuverability, or attrition); variability in target requirements and priority; variable costs (e.g., fuel, maintenance, acquisition); and a huge number of other issues that might arise when needing to coordinate a large number of heterogeneous platforms. Modeling of higher fidelity details such as these might change the performance predictions and might illustrate nuanced relationships but should not affect the findings observed here. For the current study, we stayed with three variables that could be clearly defined and implemented and found that the fitness space for heuristics is already quite complex.

Future experiments might consider a variety of explorations regarding the robustness of these findings in the context of elaborated models that represent a hybrid between the current abstraction and an actual proposed Mosaic warfare application. This model would provide a solid test bed to explore the other dimensions described earlier (e.g., heterogeneity of speed, reliability, survivability) and other relevant factors, such as the impact of jamming and the costs of greater coordination. Such elaborated models would likely require the development of more-complex MoEs that would shift the focus from peak performance to robustness-seeking Mosaic configurations and coordination strategies that perform well across a wide variety of adversary profiles.

We conclude that DARPA would be wise to invest in the development of teaming and orchestration strategies that are dynamic and adaptive, so that Mosaic platforms are afforded the ability to respond to changes and uncertainty in the environment by altering how they coordinate their activities. This would be a critical enabling technology for Mosaic warfare. Even in our simple modeling environment, it takes relatively small amounts of environmental complexity to produce a very challenging environment for rule design and selection. Additional simulations that work at this level of abstraction might produce additional insights into best approaches to system design and orchestration, but the devil is likely to remain in the details.

A second observation, resulting from our finding on the complexity of orchestration for fractionated forces, is the unprecedented challenge of demonstrating and testing such systems. The DoD's acquisition system typically relies on rigorously defined and tested requirements to guarantee the performance and behavior of an acquired system under predefined conditions. Architectures as revolutionary as those envisioned for Mosaic warfare are unlikely to be pursued unless there is ample evidence of their advantages. The dynamic nature of a Mosaic warfare system, particularly if its orchestration rules are adaptive, makes this a complex challenge. Testing for such complex adaptive systems as this one is incredibly difficult; there is simply no way to predefine all of the threat scenarios that a Mosaic system will face, and the possible permutations and behaviors of that system in response are too varied to test exhaustively.[1] In many ways, this echoes looming challenges with acquiring systems using artificial intelligence and machine learning. These systems also will be difficult to test robustly, can behave in inexplicable ways, and resist traditional techniques for formal verification.

In response to this unprecedented challenge, DARPA might wish to investigate the potential utility of new testing and acquisition strategies. New approaches could include adversarial testing (with a designated Red team attempting to defeat systems, perhaps via simulation); hard limits on the geographic, temporal, or behavioral domain that the system is allowed to operate in, with excursions treated similarly to range-safety violations; or perhaps performance-based requirements that focus on a system's impact to an overall operation, rather than individual key performance metrics. In this case, performance shortfalls or even negative performance (e.g., fratricide) would be tolerated as long as, over time and over multiple missions, the Mosaic architecture was positively contributing to an operation. This "take the bad with the good" approach could be as much of a change to defense acquisition as Mosaic warfare is to conflict in general.

---

[1] John H. Miller, and Scott E. Page, *Complex Adaptive Systems: An Introduction to Computational Models of Social Life*, Princeton, N.J.: Princeton University Press, 2007.

References

Clive, Peter D., Jeffrey A. Johnson, Michael J. Moss, James M. Zeh, Brian M. Birkmire, and Douglas D. Hodson, "Advanced Framework for Simulation, Integration and Modeling (AFSIM)," in *Proceedings from International Conference on Scientific Computing (CSC '15)*, Las Vegas, July 2015.

DARPA—*See* Defense Advanced Research Projects Agency.

Defense Advanced Research Projects Agency, "DARPA Tiles Together a Vision of Mosaic Warfare," webpage, undated. As of February 5, 2020:
https://www.darpa.mil/work-with-us/
darpa-tiles-together-a-vision-of-mosiac-warfare

Deptula, David, Heather Penney, Lawrence Stutzriem, and Mark Ginzinger, *Restoring America's Military Competitiveness: Mosaic Warfare*, Arlington, Va.: Mitchell Institute for Aerospace Studies, September 2019.

Grana, Justin, Jonathan Lamb, and Nicholas A. O'Donoughue, *Findings on Mosaic Warfare from a Colonel Blotto Game*, Santa Monica, Calif: RAND Corporation, RR-4397-OSD, 2021.

Lawler, E. L., J. K. Lenstra, A. H. G. Rinnooy Kan, and D. B. Shmoys, *The Traveling Salesman Problem: A Guided Tour of Combinatorial Optimization*, Hoboken, N.J.: John Wiley and Sons, 1985.

Lexico.com, "Monolithic," webpage, undated. As of February 5, 2020:
https://www.lexico.com/en/definition/monolithic

Miller, John H., and Scott E. Page, *Complex Adaptive Systems: An Introduction to Computational Models of Social Life*, Princeton, N.J.: Princeton University Press, 2007.

O'Donoughue, Nicholas A., Samantha McBirney, and Brian Persons, *Distributed Kill Chains: Drawing Insights for Mosaic Warfare from the Immune System and from the Navy*, Santa Monica, Calif.: RAND Corporation, RR-A573-1, 2021.

U.S. Senate, Senate Armed Services Committee, "Statement of Admiral Philip S. Davidson, U.S. Navy Commander, U.S. Indo-Pacific Command, Before the Senate Armed Services Committee on U.S. Indo-Pacific Command Posture," Washington, D.C., February 12, 2019. As of February 5, 2020: https://www.armed-services.senate.gov/imo/media/doc/Davidson_02-12-19.pdf

Wilenski, U., *NetLogo*, software package, Center for Connected Learning and Computer-Based Modeling, Northwestern University, Evanston, Ill., 1999.